Broadband Internet Access and the Digital Divide: Federal Assistance Programs

Lennard G. Kruger
Specialist in Science and Technology Policy

Angele A. Gilroy
Specialist in Telecommunications Policy

September 7, 2012

Congressional Research Service
7-5700
www.crs.gov
RL30719

CRS Report for Congress
Prepared for Members and Committees of Congress

Summary

The "digital divide" is a term that has been used to characterize a gap between "information haves and have-nots," or in other words, between those Americans who use or have access to telecommunications and information technologies and those who do not. One important subset of the digital divide debate concerns high-speed Internet access and advanced telecommunications services, also known as broadband. Broadband is provided by a series of technologies (e.g., cable, telephone wire, fiber, satellite, wireless) that give users the ability to send and receive data at volumes and speeds far greater than traditional "dial-up" Internet access over telephone lines.

Broadband technologies are currently being deployed primarily by the private sector throughout the United States. While the numbers of new broadband subscribers continue to grow, studies and data suggest that the rate of broadband deployment in urban/suburban and high income areas is outpacing deployment in rural and low-income areas. Some policymakers, believing that disparities in broadband access across American society could have adverse economic and social consequences on those left behind, assert that the federal government should play a more active role to avoid a "digital divide" in broadband access.

With the conclusion of the grant and loan awards established by the American Recovery and Reinvestment Act of 2009 (P.L. 111-5), there remain two ongoing federal vehicles which direct federal money to fund broadband infrastructure: the broadband and telecommunications programs at the Rural Utilities Service (RUS) of the U.S. Department of Agriculture and the Universal Service Fund (USF) programs under the Federal Communications Commission (FCC). Although the USF's High Cost Program does not explicitly fund broadband infrastructure, subsidies are used, in many cases, to upgrade existing telephone networks so that they are capable of delivering high-speed services. Additionally, subsidies provided by USF's Schools and Libraries Program and Rural Health Care Program are used for a variety of telecommunications services, including broadband access. Currently the USF is undergoing a major transition to the Connect America Fund, which is targeted to the deployment, adoption, and utilization of both fixed and mobile broadband.

To the extent that the 112th Congress may consider various options for further encouraging broadband deployment and adoption, a key issue is how to strike a balance between providing federal assistance for unserved and underserved areas where the private sector may not be providing acceptable levels of broadband service, while at the same time minimizing any deleterious effects that government intervention in the marketplace may have on competition and private sector investment.

Contents

Introduction 1
Status of Broadband Deployment in the United States 1
Broadband in Rural Areas 7
Are Broadband Deployment Data Adequate? 9
Broadband and the Federal Role 11
    The National Broadband Plan 11
Current Federal Broadband Programs 13
    Rural Utilities Service Programs 14
    The Universal Service Concept and the FCC 14
        Universal Service and the Telecommunications Act of 1996 14
        Universal Service and Broadband 16
Legislation in the 110th Congress 17
Legislation in the 111th Congress 17
    P.L. 111-5: The American Recovery and Reinvestment Act of 2009 18
    Other Broadband Legislation in the 111th Congress 18
Legislation in the 112th Congress 21
Concluding Observations 23

Tables

Table 1. Percentage of Broadband Technologies by Types of Connection 2
Table 2. Percentage of Households With Broadband Connections by State 2
Table 3. Americans Without Access to Fixed Broadband by State 4
Table 4. Selected Federal Domestic Assistance Programs Related to Broadband and
    Telecommunications Development 24

Contacts

Author Contact Information 29

Introduction

The "digital divide" is a term used to describe a perceived gap between "information haves and have-nots," or in other words, between those Americans who use or have access to telecommunications and information technologies and those who do not.[1] Whether or not individuals or communities fall into the "information haves" category depends on a number of factors, ranging from the presence of computers in the home, to training and education, to the availability of affordable Internet access.

Broadband technologies are currently being deployed primarily by the private sector throughout the United States. While the numbers of new broadband subscribers continue to grow, studies and data suggest that the rate of broadband deployment in urban/suburban and high income areas is outpacing deployment in rural and low-income areas.

Status of Broadband Deployment in the United States

Prior to the late 1990s, American homes accessed the Internet at maximum speeds of 56 kilobits per second by dialing up an Internet Service Provider (such as AOL) over the same copper telephone line used for traditional voice service. A relatively small number of businesses and institutions used broadband or high speed connections through the installation of special "dedicated lines" typically provided by their local telephone company. Starting in the late 1990s, cable television companies began offering cable modem broadband service to homes and businesses. This was accompanied by telephone companies beginning to offer DSL service (broadband over existing copper telephone wireline). Growth has been steep, rising from 2.8 million high speed lines reported as of December 1999, to 206 million lines as of June 30, 2011.[2] Of the 168 million high speed lines reported by the FCC, 140 million serve residential users.[3]

Table 1 depicts the relative deployment of different types of broadband technologies. A distinction is often made between "current generation" and "next generation" broadband (commonly referred to as next generation networks or NGN). "Current generation" typically refers to currently deployed cable, DSL, and many wireless systems, while "next generation" refers to dramatically faster download and upload speeds offered by fiber technologies and also by successive generations of cable, DSL, and wireless technologies.[4] In general, the greater the

---

[1] The term "digital divide" can also refer to international disparities in access to communications and information technology. This report focuses on domestic issues only.

[2] FCC, *Internet Access Services: Status as of June 30, 2011*, released June 2012, p. 16. Available at http://transition.fcc.gov/Daily_Releases/Daily_Business/2012/db0614/DOC-314630A1.pdf.

[3] Ibid.

[4] Initially, and for many years following, the FCC defined broadband (or more specifically "high-speed lines") as over 200 kilobits per second (kbps) in at least one direction, which was roughly four times the speed of conventional dialup Internet access. In recent years, the 200 kbps threshold was considered too low, and on March 19, 2008, the FCC adopted a report and order (FCC 08-89) establishing new categories of broadband speed tiers for data collection purposes. Specifically, 200 kbps to 768 kbps is considered "first generation," 768 kbps to 1.5 Mbps is "basic broadband tier 1," and increasingly higher speed tiers are broadband tiers 2 through 7 (tier seven is greater than or equal to 100 Mbps in any one direction). Tiers can change as technology advances.

download and upload speeds offered by a broadband connection, the more sophisticated (and potentially valuable) the application that is enabled.

Table 1. Percentage of Broadband Technologies by Types of Connection

	Connections over 200 kbps in at least one direction	Residential connections over 200 kbps in at least one direction	Connections at least 3 Mbps downstream and 768 kbps upstream	Residential connections at least 3 Mbps downstream and 768 kbps upstream
cable modem	22.7%	25.9%	51.8%	61.3%
DSL	15.3%	16.2%	13.5%	15.0%
Mobile wireless	58.0%	54.1%	26.2%	14.1%
Fiber	2.7%	3.0%	7.9%	9.3%
All other	1.4%	0.9%	0.6%	0.3%

Source: FCC, Internet Access Services: Status as of June 30, 2011, pp. 24-27.

Based on the latest FCC broadband connection data, **Table 2** shows the percentages of households with broadband connections by state, both for download connections over 200 kbps and for connections of at least 3 Mbps (which approximates the FCC's National Broadband Availability target). According to the FCC, high speed connections over 200 kbps are reported in 67% of households nationwide, while connections of at least 3 Mbps (download) and 768 kbps (upload) are reported in 38% of households nationwide.

Table 2. Percentage of Households With Broadband Connections by State
(as of June 30, 2011)

	Connections over 200 kbps	Connections at least 3 mbps downstream and 768 kbps upstream
Alabama	56%	23%
Alaska	70%	*
Arizona	65%	40%
Arkansas	52%	19%
California	74%	44%
Colorado	73%	53%
Connecticut	78%	51%
Delaware	76%	66%
District of Columbia	68%	55%
Florida	72%	42%
Georgia	62%	34%
Hawaii	*	*
Idaho	60%	17%

Broadband Internet Access and the Digital Divide: Federal Assistance Programs

	Connections over 200 kbps	Connections at least 3 mbps downstream and 768 kbps upstream
Illinois	66%	35%
Indiana	62%	33%
Iowa	65%	21%
Kansas	66%	25%
Kentucky	59%	33%
Louisiana	58%	28%
Maine	74%	22%
Maryland	74%	65%
Massachusetts	78%	69%
Michigan	64%	38%
Minnesota	68%	41%
Mississippi	46%	13%
Missouri	60%	22%
Montana	65%	35%
Nebraska	67%	41%
Nevada	64%	36%
New Hampshire	79%	54%
New Jersey	80%	73%
New Mexico	59%	32%
New York	73%	48%
North Carolina	66%	13%
North Dakota	68%	35%
Ohio	66%	19%
Oklahoma	57%	24%
Oregon	67%	48%
Pennsylvania	70%	51%
Rhode Island	74%	*
South Carolina	60%	20%
South Dakota	61%	37%
Tennessee	55%	31%
Texas	63%	28%
Utah	70%	47%
Vermont	74%	54%
Virginia	68%	57%
Washington	71%	53%
West Virginia	57%	27%

	Connections over 200 kbps	Connections at least 3 mbps downstream and 768 kbps upstream
Wisconsin	67%	25%
Wyoming	64%	43%
National subscribership ratio	67%	38%

Source: FCC, *Internet Access Services: Status as of June 30, 2011*, pp. 35-36.

Notes: Asterisk (*) indicates data withheld by the FCC to maintain firm confidentiality. Subscribership ratio is the number of reported residential high speed lines (broadband connections) divided by the number of households in each state.

Meanwhile, the National Broadband Map, which is composed of state broadband data and compiled by NTIA, provides data on where broadband is and is not available. The latest update of these data indicate that 99.6% of the U.S. population has minimum advertised broadband speeds available (at least 768 kbps download/200 kbps upload), while 96.7% have available advertised speeds of at least 3 Mbps (download) and 768 kbps (upload).[5] The FCC's *Eighth Broadband Progress Report*, released on August 21, 2012, used National Broadband Map data to estimate that 19 million Americans living in 7 million households lack access to fixed broadband at speeds of 4 Mbps (download)/1 Mbps (upload) or greater.[6] **Table 3** shows a state-by-state breakdown of the percentage of population without access to fixed broadband at the FCC's benchmark speed of 4 Mbps/1Mbps.

Table 3. Americans Without Access to Fixed Broadband by State

(access to speeds of at least 4 Mbps download/1 Mbps upload)

	% of population without access	% of population without access, nonrural areas	% of population without access, rural areas
United States	6.0%	1.8%	23.7%
Alabama	11.4%	1.6%	25.5%
Alaska	19.6%	4.4%	48.9%
Arizona	4.7%	1.2%	35.8%
Arkansas	13.6%	1.8%	28.8%
California	3.3%	1.6%	35.2%
Colorado	4.3%	1.0%	25.3%
Connecticut	0.7%	0.5%	2.6%
Delaware	3.1%	1.1%	13.0%
District of Columbia	0.0%	0.0%	N/A
Florida	3.1%	2.0%	14.3%
Georgia	3.4%	1.3%	9.9%

---

[5] Data as of December 31, 2011. Available at http://www.broadbandmap.gov/summarize/nationwide.

[6] Federal Communications Commission, *Eighth Broadband Progress Report*, FCC 12-90, released August 21, 2012, p. 29, available at http://transition.fcc.gov/Daily_Releases/Daily_Business/2012/db0827/FCC-12-90A1.pdf.

	% of population without access	% of population without access, nonrural areas	% of population without access, rural areas
Hawaii	1.5%	0.1%	17.7%
Idaho	13.1%	1.3%	41.4%
Illinois	3.3%	0.4%	25.6%
Indiana	4.3%	1.3%	12.4%
Iowa	7.1%	0.7%	18.7%
Kansas	7.7%	1.0%	27.0%
Kentucky	10.5%	1.5%	23.0%
Louisiana	8.8%	1.3%	29.6%
Maine	4.7%	1.2%	7.0%
Maryland	3.2%	0.9%	19.2%
Massachusetts	1.0%	0.5%	6.4%
Michigan	6.3%	0.8%	22.4%
Minnesota	8.0%	0.8%	27.7%
Mississippi	12.1%	1.2%	22.8%
Missouri	7.5%	0.6%	24.2%
Montana	26.7%	4.0%	55.4%
Nebraska	10.1%	1.9%	33.0%
Nevada	2.3%	0.6%	30.2%
New Hampshire	7.5%	2.5%	15.2%
New Jersey	0.7%	0.4%	5.6%
New Mexico	14.2%	4.8%	46.7%
New York	1.3%	0.0%	10.4%
North Carolina	6.4%	2.1%	15.0%
North Dakota	15.9%	2.5%	36.2%
Ohio	3.4%	0.5%	14.0%
Oklahoma	16.2%	2.9%	42.5%
Oregon	3.4%	0.2%	17.3%
Pennsylvania	1.7%	0.3%	6.8%
Rhode Island	0.2%	0.0%	2.3%
South Carolina	11.7%	4.9%	25.1%
South Dakota	21.1%	3.2%	44.6%
Tennessee	6.8%	0.9%	18.6%
Texas	5.9%	2.0%	27.6%
Utah	1.8%	0.3%	16.7%
Vermont	9.4%	0.2%	15.2%
Virginia	10.9%	2.2%	37.6%

	% of population without access	% of population without access, nonrural areas	% of population without access, rural areas
Washington	3.2%	0.5%	17.4%
West Virginia	45.9%	31.4%	59.8%
Wisconsin	6.9%	0.1%	23.0%
Wyoming	13.2%	1.1%	35.4%
U.S. Territories	**54.0%**	**41.5%**	**85.2%**
American Samoa	78.6%	30.9%	92.0%
Northern Marianas	100.0%	100.0%	100.0%
Guam	54.3%	0.1%	76.1%
Puerto Rico	51.6%	40.3%	84.8%
U.S. Virgin Islands	100.0%	100.0%	100.0%

Source: FCC, *Eighth Broadband Progress Report*, Appendix C.

In contrast to broadband *availability*, which refers to whether or not broadband service is offered, broadband *adoption* refers to the extent to which American households actually subscribe to and use broadband. The U.S. Department of Commerce report, *Exploring the Digital Nation: Computer and Internet Use at Home* (based on October 2010 U.S. Census Bureau survey data) found that 68% of U.S. households have adopted broadband.[7] Similarly, the FCC's *Eighth Broadband Progress Report* found that 64% of American households with broadband available to them adopt broadband service offering speeds faster than 768 kbps/200 kbps, while 40% adopt speeds faster than the FCC benchmark of 4 Mbps/1Mbps. The FCC found that the "broadband adoption rates for American households are lower, on average, in the counties with the lowest median household income, in areas outside of urban areas, on Tribal lands, and in U.S. Territories."[8]

An FCC consumer survey, conducted in October and November 2009, found that 35% or 80 million American adults do not use broadband at home, falling into three categories: those who do not use the Internet at all (22%); those who use the Internet but do not have Internet access at home (6%); and those who use dial-up to access the Internet (6%). The survey identified three primary reasons why non-adopting Americans do not have broadband: cost, lack of digital literacy, and the perceived insufficient relevance of broadband.[9] Similarly, according to the Department of Commerce report, the two most common reasons cited for not having broadband at home are that it is perceived as not needed or too expensive. Lack of a home computer can also be a major factor.[10] The Department of Commerce report, the FCC's National Broadband Plan,

---

[7] U.S. Department of Commerce, National Telecommunications and Information Administration, *Exploring the Digital Nation: Computer and Internet Use at Home*, November 2011, p. v, available at http://www.ntia.doc.gov/files/ntia/publications/exploring_the_digital_nation_computer_and_internet_use_at_home_11092011.pdf.

[8] *Eighth Broadband Progress Report*, p. 54.

[9] Horrigan, John, Federal Communications Commission, *Broadband Adoption and Use in America*, OBI Working Paper Series No. 1, February 2010, p. 5, available at http://hraunfoss.fcc.gov/edocs_public/attachmatch/DOC-296442A1.pdf.

[10] *Digital Nation*, p. 20.

and a survey conducted by the Pew Internet and American Life Project[11] also found disparities in broadband adoption among demographic groups. Populations continuing to lag behind in broadband adoption include people with low incomes, seniors, minorities, the less-educated, non-family households, and the non-employed.

Broadband in Rural Areas[12]

While the number of new broadband subscribers continues to grow, the rate of broadband deployment in urban areas appears to be outpacing deployment in rural areas. While there are many examples of rural communities with state of the art telecommunications facilities,[13] recent surveys and studies have indicated that, in general, rural areas tend to lag behind urban and suburban areas in broadband deployment. For example:

- According to the FCC's *Eighth Broadband Progress Report*, of the 19 million Americans who live where fixed broadband is unavailable, 14.5 million live in rural areas.[14]

- The Department of Commerce report, *Exploring the Digital Nation*, found that while the digital divide between urban and rural areas has lessened since 2007, it still persists with 70% of urban households adopting broadband service in 2010, compared to 57% of rural households.[15]

- Data from the Pew Internet & American Life Project indicate that while broadband adoption is growing in rural areas, broadband users make up larger percentages of non-rural users than rural users. Pew found that the percentage of all U.S. adults with broadband at home is 70% for non-rural areas and 50% for rural areas.[16]

- According to December 2011 data from the National Broadband Map, 99.7% of the population in urban areas have access to available broadband speeds of at least 3 Mbps (download)/768 kbps (upload), as opposed to 84.0% of the population in rural areas.[17]

The comparatively lower population density of rural areas is likely the major reason why broadband is less deployed than in more highly populated suburban and urban areas. Particularly for wireline broadband technologies—such as cable modem and DSL—the greater the geographical distances among customers, the larger the cost to serve those customers. Thus, there is often less incentive for companies to invest in broadband in rural areas than, for example, in an

---

[11] Smith, Aaron, Pew Internet & American Life Project, *Home Broadband 2010*, August 11, 2010, available at http://www.pewinternet.org/~/media//Files/Reports/2010/Home%20broadband%202010.pdf.

[12] For more information on rural broadband and broadband programs at the Rural Utilities Service, see CRS Report RL33816, *Broadband Loan and Grant Programs in the USDA's Rural Utilities Service*, by Lennard G. Kruger.

[13] See for example: National Exchange Carrier Association (NECA), Trends 2006: Making Progress With Broadband, 2006, 26 p. Available at http://www.neca.org/media/trends_brochure_website.pdf.

[14] *Eighth Broadband Progress Report*, p. 5.

[15] *Exploring the Digital Nation*, p. 24.

[16] *Home Broadband 2010*, p. 8.

[17] NTIA, National Broadband Map, *Broadband Statistics Report: Broadband Availability in Urban vs. Rural Areas*, p. 7.

urban area where there is more demand (more customers with perhaps higher incomes) and less cost to wire the market area.[18]

Some policymakers believe that disparities in broadband access across American society could have adverse consequences on those left behind, and that advanced telecommunications applications critical for businesses and consumers to engage in e-commerce are increasingly dependent on high speed broadband connections to the Internet. Thus, some say, communities and individuals without access to broadband could be at risk to the extent that connectivity becomes a critical factor in determining future economic development and prosperity. A February 2006 study done by the Massachusetts Institute of Technology for the Economic Development Administration of the Department of Commerce marked the first attempt to quantitatively measure the impact of broadband on economic growth. The study found that "between 1998 and 2002, communities in which mass-market broadband was available by December 1999 experienced more rapid growth in employment, the number of businesses overall, and businesses in IT-intensive sectors, relative to comparable communities without broadband at that time."[19]

A June 2007 report from the Brookings Institution found that for every one percentage point increase in broadband penetration in a state, employment is projected to increase by 0.2% to 0.3% per year. For the entire U.S. private non-farm economy, the study projected an increase of about 300,000 jobs.[20]

Subsequently, a July 2009 study commissioned by the Internet Innovation Alliance found net consumer benefits of home broadband on the order of $32 billion per year, up from an estimated $20 billion in consumer benefits from home broadband in 2005.[21]

Some also argue that broadband is an important contributor to U.S. future economic strength with respect to the rest of the world. Data from the Organization for Economic Cooperation and Development (OECD) found the U.S. ranking 15th among OECD nations in broadband access per 100 inhabitants as of December 2011.[22] By contrast, in 2001 an OECD study found the U.S. ranking fourth in broadband subscribership per 100 inhabitants (after Korea, Sweden, and Canada).[23] While many argue that declining U.S. performance in international broadband rankings is a cause for concern,[24] others maintain that the OECD data undercount U.S. broadband

---

[18] The terrain of rural areas can also be a hindrance to broadband deployment because it is more expensive to deploy broadband technologies in a mountainous or heavily forested area. An additional added cost factor for remote areas can be the expense of "backhaul" (e.g., the "middle mile") which refers to the installation of a dedicated line which transmits a signal to and from an Internet backbone which is typically located in or near an urban area.

[19] Gillett, Sharon E., Massachusetts Institute of Technology, *Measuring Broadband's Economic Impact*, report prepared for the Economic Development Administration, U.S. Department of Commerce, February 28, 2006, p. 4.

[20] Crandall, Robert, William Lehr, and Robert Litan, *The Effects of Broadband Deployment on Output and Employment: A Cross-sectional Analysis of U.S. Data*, June 2007, 20 pp. Available at http://www3.brookings.edu/views/papers/crandall/200706litan.pdf.

[21] Mark Dutz, Jonathan Orszag, and Robert Willig, *The Substantial Consumer Benefits of Broadband Connectivity for U.S. Households*, Internet Innovation Alliance, July 2009, p. 4, http://internetinnovation.org/files/special-reports/CONSUMER_BENEFITS_OF_BROADBAND.pdf.

[22] OECD, *OECD Broadband Portal*. Available at http://www.oecd.org/sti/ict/broadband.

[23] OECD, Directorate for Science, Technology and Industry, *The Development of Broadband Access in OECD Countries*, October 29, 2001, 63 pp. For a comparison of government broadband policies, also see OECD, Directorate for Science, Technology and Industry, *Broadband Infrastructure Deployment: The Role of Government Assistance*, May 22, 2002, 42 pp.

[24] See Turner, Derek S., Free Press, *Broadband Reality Check II: The Truth Behind America's Digital Divide*, August (continued...)

deployment,[25] and that cross-country broadband deployment comparisons are not necessarily meaningful and are inherently problematic.[26] Finally, an issue related to international broadband rankings is the extent to which broadband speeds and prices differ between the United States and the rest of the world.[27]

Are Broadband Deployment Data Adequate?

Obtaining an accurate snapshot of the status of broadband deployment is problematic. Anecdotes abound of rural and low-income areas which do not have adequate Internet access, as well as those which are receiving access to high-speed, state-of-the-art connections. Rapidly evolving technologies, the constant flux of the telecommunications industry, the uncertainty of consumer wants and needs, and the sheer diversity and size of the nation's economy and geography make the status of broadband deployment very difficult to characterize. The FCC periodically collects broadband deployment data from the private sector via "FCC Form 477"—a standardized information gathering survey. Statistics derived from the Form 477 survey are published every six months. Additionally, data from Form 477 are used as the basis of the FCC's (to date) eight broadband deployment reports.

The FCC is working to refine the data used in future reports in order to provide an increasingly accurate portrayal. In its March 17, 2004, Notice of Inquiry for the *Fourth Report*, the FCC sought comments on specific proposals to improve the FCC Form 477 data gathering program.[28] On November 9, 2004, the FCC voted to expand its data collection program by requiring reports from all facilities based carriers regardless of size in order to better track rural and underserved markets, by requiring broadband providers to provide more information on the speed and nature of their service, and by establishing broadband-over-power line as a separate category in order to track its development and deployment. The FCC Form 477 data gathering program was extended for five years beyond its March 2005 expiration date.[29]

(...continued)
2006, pp 8-11. Available at http://www.freepress.net/files/bbrc2-final.pdf; and Turner, Derek S., Free Press, *'Shooting the Messenger' Myth vs. Reality: U.S. Broadband Policy and International Broadband Rankings*, July 2007, 25 pp., available at http://www.freepress.net/files/shooting_the_messenger.pdf.

[25] National Telecommunications and Information Administration, *Fact Sheet: United States Maintains Information and Communication Technology (ICT) Leadership and Economic Strength*, at http://www.ntia.doc.gov/ntiahome/press/2007/ICTleader_042407.html.

[26] See Wallsten, Scott, Progress and Freedom Foundation, *Towards Effective U.S. Broadband Policies*, May 2007, 19 pp. Available at http://www.pff.org/issues-pubs/pops/pop14.7usbroadbandpolicy.pdf. Also see Ford, George, Phoenix Center, *The Broadband Performance Index: What Really Drives Broadband Adoption Across the OECD?*, Phoenix Center Policy Paper Number 33, May 2008, 27 pp; available at http://www.phoenix-center.org/pcpp/PCPP33Final.pdf.

[27] See price and services and speed data on OECD Broadband Portal, available at http://www.oecd.org/sti/ict/broadband; see also Federal Communications Commission, *International Broadband Data Report*, IB Docket No. 10-171, DA 11-732, May 20, 2011, available at http://hraunfoss.fcc.gov/edocs_public/attachmatch/DA-11-732A1.pdf.

[28] Federal Communications Commission, *Notice of Inquiry*, "Concerning the Deployment of Advanced Telecommunications Capability to All Americans in a Reasonable and Timely Fashion, and possible Steps to Accelerate Such Deployment Pursuant to Section 706 of the Telecommunications Act of 1996," FCC 04-55, March 17, 2004, p. 6.

[29] FCC News Release, *FCC Improves Data Collection to Monitor Nationwide Broadband Rollout*, November 9, 2004. Available at http://hraunfoss.fcc.gov/edocs_public/attachmatch/DOC-254115A1.pdf.

On April 16, 2007, the FCC announced a Notice of Proposed Rulemaking which sought comment on a number of broadband data collection issues, including how to develop a more accurate picture of broadband deployment; gathering information on price, other factors determining consumer uptake of broadband, and international comparisons; how to improve data on wireless broadband; how to collect information on subscribership to voice over Internet Protocol service (VoIP); and whether to modify collection of speed tier information.[30]

On March 19, 2008, the FCC adopted an order that substantially expands its broadband data collection capability. Specifically, the order expands the number of broadband reporting speed tiers to capture more information about upload and download speeds offered in the marketplace, requires broadband providers to report numbers of broadband subscribers by census tract, and improves the accuracy of information collected on mobile wireless broadband deployment. Additionally, in a Further Notice of Proposed Rulemaking, the FCC sought comment on broadband service pricing and availability.[31] The July 2009 data release (providing data as of June 30, 2008) was the final data set gathered under the old FCC Form 477. The February 2010 data report (December 31, 2008, data) reflected the new Form 477 data collection requirements.

Meanwhile, during the 110th Congress, state initiatives to collect broadband deployment data in order to promote broadband in underserved areas were viewed as a possible model for governmental efforts to encourage broadband. The Broadband Data Improvement Act was enacted by the 110th Congress and became P.L. 110-385 on October 10, 2008. The law requires the FCC to collect demographic information on unserved areas, data comparing broadband service with 75 communities in at least 25 nations abroad, and data on consumer use of broadband. The act also directs the Census Bureau to collect broadband data, the Government Accountability Office to study broadband data metrics and standards, and the Department of Commerce to provide grants supporting state broadband initiatives.

P.L. 111-5, the American Recovery and Reinvestment Act, provided NTIA with an appropriation of $350 million to implement the Broadband Data Improvement Act and to develop and maintain a national broadband inventory map. The National Broadband Map was released on February 17, 2011 (http://www.broadbandmap.gov), and will be updated every six months.[32]

Finally, the FCC's National Broadband Plan addressed the broadband data issue, recommending that the FCC and the U.S. Bureau of Labor Statistics (BLS) should collect more detailed and accurate data on actual availability, penetration, prices, churn, and bundles offered by broadband service providers to consumers and businesses, and should publish analyses of these data.

---

[30] Federal Communications Commission, *Notice Proposed Rulemaking*, "Development of Nationwide Broadband Data to Evaluate Reasonable and Timely Deployment of Advanced Services to All Americans, Improvement of Wireless Broadband Subscribership Data, and Development of Data on Interconnected Voice Over Internet Protocol (VoIP) Subscribership," WC Docket No. 07-38, FCC 07-17, released April 16, 2007, 56 pp.

[31] FCC, News Release, "FCC Expands, Improves Broadband Data Collection," March 19, 2008. Available at http://hraunfoss.fcc.gov/edocs_public/attachmatch/DOC-280909A1.pdf.

[32] For more information on the national broadband mapping program and the State Broadband Data and Development Program, see http://www.ntia.doc.gov/broadbandgrants/broadbandmapping.html.

Broadband and the Federal Role

The Telecommunications Act of 1996 (P.L. 104-104) addressed the issue of whether the federal government should intervene to prevent a "digital divide" in broadband access. Section 706 requires the FCC to determine whether "advanced telecommunications capability [i.e., broadband or high-speed access] is being deployed to all Americans in a reasonable and timely fashion."

Since 1999, the FCC has adopted and released eight reports pursuant to Section 706. The first five reports formally concluded that the deployment of advanced telecommunications capability to all Americans is reasonable and timely. Unlike the first five 706 reports, the sixth, seventh, and eighth reports concluded that broadband is not being deployed to all Americans in a reasonable and timely fashion. According to the *Eighth Broadband Progress Report*:

> Our analysis shows that the nation's broadband deployment gap remains significant and is particularly pronounced for Americans living in rural areas and on Tribal lands. We find that as of June 30, 2011, approximately 19 million Americans did not have access to fixed broadband. Significantly, approximately 76 percent of these Americans reside in rural areas. Our analysis further shows that Americans residing on Tribal lands disproportionately lack access to fixed broadband. And the available international broadband data, though not perfectly comparable to U.S. data, suggest that the availability and deployment of broadband in the United States may lag behind a number of other developed countries in certain respects, although we also compare favorably to some developed countries in other respects. Moreover, as many as 80 percent of E-rate recipients say that their broadband connections do not fully meet their needs, and 78 percent of recipients say that they need additional bandwidth. These data combined with our findings concerning availability above provide further indication that broadband is not yet being reasonably and timely deployed to all Americans.[33]

FCC Commissioners Robert McDowell and Ajit Pai issued dissenting statements, maintaining that there is insufficient justification for the 706 report conclusion that broadband is not being deployed in a reasonable and timely fashion. For example, the dissents argued that the report did not sufficiently account for the dramatic growth in the availability and deployment of mobile broadband, and that gaps in broadband adoption should not be used to determine whether or not broadband is being sufficiently deployed.[34]

The National Broadband Plan

As mandated by the ARRA, on March 16, 2010, the FCC publically released its report, *Connecting America: The National Broadband Plan*.[35] The National Broadband Plan (NBP) seeks to "create a high-performance America," which the FCC defines as "a more productive, creative, efficient America in which affordable broadband is available everywhere and everyone has the means and skills to use valuable broadband applications."[36] In order to achieve this mission, the NBP recommends that the country set six goals for 2020:

---

[33] *Eighth Broadband Progress Report*, p. 59-60.

[34] Ibid., p.171, 177.

[35] Available at http://www.broadband.gov/plan/. For more information on the National Broadband Plan, see CRS Report R41324, *The National Broadband Plan*, by Lennard G. Kruger et al.

[36] Federal Communications Commission, Connecting America: *The National Broadband Plan*, March 17, 2010, p. 9.

- Goal No. 1: At least 100 million U.S. homes should have affordable access to actual download speeds of at least 100 megabits per second and actual upload speeds of at least 50 megabits per second.

- Goal No. 2: The United States should lead the world in mobile innovation, with the fastest and most extensive wireless networks of any nation.

- Goal No. 3: Every American should have affordable access to robust broadband service, and the means and skills to subscribe if they so choose.

- Goal No. 4: Every American community should have affordable access to at least 1 gigabit per second broadband service to anchor institutions such as schools, hospitals, and government buildings.

- Goal No. 5: To ensure the safety of the American people, every first responder should have access to a nationwide, wireless, interoperable broadband public safety network.

- Goal No. 6: To ensure that America leads in the clean energy economy, every American should be able to use broadband to track and manage their real-time energy consumption.

The National Broadband Plan is categorized into three parts:

- **Part I (Innovation and Investment)**, which "discusses recommendations to maximize innovation, investment and consumer welfare, primarily through competition. It then recommends more efficient allocation and management of assets government controls or influences."[37] The recommendations address a number of issues, including spectrum policy, improved broadband data collection, broadband performance standards and disclosure, special access rates, interconnection, privacy and cybersecurity, child online safety, poles and rights-of-way, research and experimentation (R&E) tax credits, and R&D funding.

- **Part II (Inclusion)**, which "makes recommendations to promote inclusion—to ensure that all Americans have access to the opportunities broadband can provide."[38] Issues include reforming the Universal Service Fund, intercarrier compensation, federal assistance for broadband in Tribal lands, expanding existing broadband grant and loan programs at the Rural Utilities Service, enabling greater broadband connectivity in anchor institutions, and improved broadband adoption and utilization especially among disadvantaged and vulnerable populations.

- **Part III (National Purposes)**, which "makes recommendations to maximize the use of broadband to address national priorities. This includes reforming laws, policies and incentives to maximize the benefits of broadband in areas where government plays a significant role."[39] National purposes include health care, education, energy and the environment, government performance, civic engagement, and public safety. Issues include telehealth and health IT, online

---

[37] Ibid., p. 11.
[38] Ibid.
[39] Ibid.

learning and modernizing educational broadband infrastructure, digital literacy and job training, smart grid and smart buildings, federal support for broadband in small businesses, telework within the federal government, cybersecurity and protection of critical broadband infrastructure, copyright of public digital media, interoperable public safety communications, next generation 911 networks, and emergency alert systems.

The release of the National Broadband Plan is seen by many as a precursor towards the development of a national broadband policy—whether comprehensive or piecemeal—that will likely be shaped and developed by Congress, the FCC, and the Administration. Upon release of the NBP, President Obama issued the following statement:

> My Administration will build upon our efforts over the past year to make America's nationwide broadband infrastructure the world's most powerful platform for economic growth and prosperity, including improving access to mobile broadband, maximizing technology innovation, and supporting a nationwide, interoperable public safety wireless broadband network.[40]

Meanwhile, Congress will play a major role in implementing the National Broadband Plan, both by considering legislation to implement NBP recommendations, and by overseeing broadband activities conducted by the FCC and executive branch agencies.

Current Federal Broadband Programs

With the conclusion of grant and loan awards established by the American Recovery and Reinvestment Act of 2009 (P.L. 111-5),[41] there remain two ongoing federal vehicles which direct federal money to fund broadband infrastructure: the broadband and telecommunications programs at the Rural Utilities Service (RUS) of the U.S. Department of Agriculture and the Universal Service Fund (USF) programs under the Federal Communications Commission (FCC). Although the USF's High Cost Program does not explicitly fund broadband infrastructure, subsidies are used, in many cases, to upgrade existing telephone networks so that they are capable of delivering high-speed services. Additionally, subsidies provided by USF's Schools and Libraries Program and Rural Health Care Program are used for a variety of telecommunications services, including broadband access. Currently the USF is undergoing a major transition to the Connect America Fund, which is targeted to the deployment, adoption, and use of both fixed and mobile broadband.

Table 4 (at the end of this report) shows selected federal domestic assistance programs throughout the federal government that currently can be associated with broadband and telecommunications development. The table categorizes the programs in three ways: programs exclusively devoted to the deployment of broadband infrastructure; programs which focus on or include deployment of telecommunications infrastructure generally (which typically can and does include broadband); and applications-specific programs which fund some aspect of broadband access or adoption as a means towards supporting a particular application, such as distance learning or telemedicine.

---

[40] The White House, Office of the Press Secretary, "Statement from the President on the National Broadband Plan," March 16, 2010, available at http://www.whitehouse.gov/the-press-office/statement-president-national-broadband-plan.

[41] See CRS Report R40436, *Broadband Infrastructure Programs in the American Recovery and Reinvestment Act*, by Lennard G. Kruger.

Rural Utilities Service Programs

RUS implements two programs specifically targeted at providing assistance for broadband infrastructure deployment in rural areas: the Rural Broadband Access Loan and Loan Guarantee Program and Community Connect Broadband Grants.[42] The 110th Congress reauthorized and reformed the Rural Broadband Access Loan and Loan Guarantee program as part of the 2008 farm bill (P.L. 110-234). The 112th Congress is considering reauthorization of the program as part of the 2012 farm bill.[43]

RUS also has a rural telephone loan program (dating back to 1949, now called Telecommunications Infrastructure Loans) that has historically supported infrastructure for telephone voice service, but has now evolved into support for broadband-capable service provided by traditional telephone borrowers. Additionally, the Distance Learning and Telemedicine Grant Program supports broadband-based applications.[44]

The Universal Service Concept and the FCC[45]

Since its creation in 1934 the Federal Communications Commission (FCC) has been tasked with "mak[ing] available, so far as possible, to all the people of the United States ... a rapid, efficient, Nation-wide, and world-wide wire and radio communications service with adequate facilities at reasonable charges."[46] This mandate led to the development of what has come to be known as the universal service concept.

The universal service concept, as originally designed, called for the establishment of policies to ensure that telecommunications services are available to all Americans, including those in rural, insular and high cost areas, by ensuring that rates remain affordable. Over the years this concept fostered the development of various FCC policies and programs to meet this goal. The FCC offers universal service support through a number of direct mechanisms that target both providers of and subscribers to telecommunications services.[47]

Universal Service and the Telecommunications Act of 1996

Passage of the Telecommunications Act of 1996 (P.L. 104-104) codified the long-standing commitment by U.S. policymakers to ensure universal service in the provision of telecommunications services.

---

[42] For more information on these programs, see CRS Report RL33816, *Broadband Loan and Grant Programs in the USDA's Rural Utilities Service*, by Lennard G. Kruger.

[43] Ibid.

[44] See CRS Report R42524, *Rural Broadband: The Roles of the Rural Utilities Service and the Universal Service Fund*, by Angele A. Gilroy and Lennard G. Kruger.

[45] The section on universal service was prepared by Angele Gilroy, Specialist in Telecommunications, Resources, Science and Industry Division. For more information on universal service, see CRS Report RL33979, *Universal Service Fund: Background and Options for Reform*, by Angele A. Gilroy.

[46] Communications Act of 1934, As Amended, Title I §1 [47 U.S.C. 151].

[47] Many states participate in or have programs that mirror FCC universal service mechanisms to help promote universal service goals within their states.

The Schools and Libraries, and Rural Health Care Programs

Congress, through the 1996 act, not only codified, but also expanded the concept of universal service to include, among other principles, that elementary and secondary schools and classrooms, libraries, and rural health care providers have access to telecommunications services for specific purposes at discounted rates. (See §§254(b)(6) and 254(h)of the 1996 Telecommunications Act, 47 U.S.C. 254.)

1. The Schools and Libraries Program. Under universal service provisions contained in the 1996 act, elementary and secondary schools and classrooms and libraries are designated as beneficiaries of universal service discounts. Universal service principles detailed in Section 254(b)(6) state that "Elementary and secondary schools and classrooms ... and libraries should have access to advanced telecommunications services." The act further requires in Section 254(h)(1)(B) that services within the definition of universal service be provided to elementary and secondary schools and libraries for education purposes at discounts, that is at "rates less than the amounts charged for similar services to other parties."

The FCC established the Schools and Libraries Division within the Universal Service Administrative Company (USAC) to administer the schools and libraries or "E (education)-rate" program to comply with these provisions. Under this program, eligible schools and libraries receive discounts ranging from 20% to 90% for telecommunications services depending on the poverty level of the school's (or school district's) population and its location in a high cost telecommunications area. The FCC established a the funding ceiling, or cap, of $2.25 billion, adjusted for inflation prospectively beginning with funding year 2010. Three categories of services are eligible for discounts: internal connections (e.g., wiring, routers and servers); Internet access; and telecommunications and dedicated services, with the third category receiving funding priority. According to data released by program administrators, approximately $31 billion in funding has been committed over the first 14 years of the program with funding released to all states, the District of Columbia and all territories. Funding commitments for funding Year 2012 (July 1, 2012-June 30, 2013), the 15th and current year of the program, totaled $863.7 million as of July 31, 2012.[48]

2. The Rural Health Care Program. Section 254(h) of the 1996 act requires that public and non-profit rural health care providers have access to telecommunications services necessary for the provision of health care services at rates comparable to those paid for similar services in urban areas. Subsection 254(h)(1) further specifies that "to the extent technically feasible and economically reasonable" health care providers should have access to advanced telecommunications and information services. The FCC established the Rural Health Care Division (RHCD) within the USAC to administer the universal support program to comply with these provisions. Under FCC established rules only public or non-profit health care providers are eligible to receive funding. Eligible health care providers, with the exception of those requesting only access to the Internet, must also be located in a rural area. The funding ceiling, or cap, for this support was established at $400 million annually. The funding level for Year One of the program (January 1998-June 30, 1999) was set at $100 million. Due to less than anticipated demand, the FCC established a $12 million funding level for the second year (July 1, 1999 to June 30, 2000) of the program but has since returned to a $400 million yearly cap. As of March

---
[48] For additional information on this program, including funding commitments, see the E-rate website: http://www.universalservice.org/sl/.

31, 2012, a total of $514.3 million has been committed since the program's inception in 1998. The primary use of the funding is to provide reduced rates for telecommunications and information services necessary for the provision of health care.[49] In addition, the FCC established, in 2007, the "Rural Health Care Pilot Program" to help public and non-profit health care providers build state and region-wide broadband networks dedicated to the provision of health care services. There are 50 projects in the program with $387.9 million in authorized funds. As of February 29, 2010, $232.6 million of the funds have been committed to the 50 FCC designated projects.

Universal Service and Broadband

One of the policy debates surrounding universal service is whether access to advanced telecommunications services (i.e., broadband) should be incorporated into universal service objectives. The term universal service, when applied to telecommunications, refers to the ability to make available a basket of telecommunications services to the public, across the nation, at a reasonable price. As directed in the 1996 Telecommunications Act (§254[c]) a federal-state Joint Board was tasked with defining the services which should be included in the basket of services to be eligible for federal universal service support; in effect using and defining the term "universal service" for the first time. The Joint Board's recommendation, which was subsequently adopted by the FCC in May 1997, included the following in its universal service package: voice grade access to and some usage of the public switched network; single line service; dual tone signaling; access to directory assistance; emergency service such as 911; operator services; and access and interexchange (long distance) service.

Some policy makers expressed concern that the FCC-adopted definition is too limited and does not take into consideration the importance and growing acceptance of advanced services such as broadband and Internet access. They point to a number of provisions contained in the Universal Service section of the 1996 act to support their claim. Universal service principles contained in Section 254(b)(2) state that "Access to advanced telecommunications services should be provided to all regions of the Nation." The subsequent principle (b)(3) calls for consumers in all regions of the nation including "low-income" and those in "rural, insular, and high cost areas" to have access to telecommunications and information services including "advanced services" at a comparable level and a comparable rate charged for similar services in urban areas. Such provisions, they state, dictate that the FCC expand its universal service definition.

The 1996 act does take into consideration the changing nature of the telecommunications sector and allows for the universal service definition to be modified if future conditions warrant. Section 254(c)of the act states that "universal service is an evolving level of telecommunications services" and the FCC is tasked with "periodically" reevaluating this definition "taking into account advances in telecommunications and information technologies and services." Furthermore, the Joint Board is given specific authority to recommend "from time to time" to the FCC modification in the definition of the services to be included for federal universal service support. The Joint Board, on November 19, 2007, concluded such an inquiry and recommended that the FCC change the mix of services eligible for universal service support. The Joint Board recommended, among other things, that "the universal availability of broadband Internet services" be included in the nation's communications goals and hence be supported by federal

---

[49] For additional information on this program, including funding commitments, see the RHCD website: http://www.universalservice.org/rhc/.

universal service funds.[50] The FCC in its national broadband plan, *Connecting America: the National Broadband Plan,* recommended that access to and adoption of broadband be a national goal. Furthermore the national broadband plan proposed that the Universal Service Fund be restructured to become a vehicle to help reach this goal. The FCC, in an October 2011 decision, adopted an Order that calls for the USF to be transformed, in stages, over a multi-year period, from a mechanism to support voice telephone service to one that supports the deployment, adoption, and utilization of both fixed and mobile broadband.[51]

Legislation in the 110th Congress

In the 110th Congress, legislation was enacted to provide financial assistance for broadband deployment. Of particular note is the reauthorization of the Rural Utilities Service (RUS) broadband loan program, which was enacted as part of the 2008 farm bill (P.L. 110-234). In addition to reauthorizing and reforming the RUS broadband loan program, P.L. 110-234 contains provisions establishing a National Center for Rural Telecommunications Assessment and requiring the FCC and RUS to formulate a comprehensive rural broadband strategy.

The Broadband Data Improvement Act (P.L. 110-385) was enacted by the 110th Congress and required the FCC to collect demographic information on unserved areas, data comparing broadband service with 75 communities in at least 25 nations abroad, and data on consumer use of broadband. The act also directed the Census Bureau to collect broadband data, the Government Accountability Office to study broadband data metrics and standards, and the Department of Commerce to provide grants supporting state broadband initiatives.

Meanwhile, the America COMPETES Act (H.R. 2272) was enacted (P.L. 110-69) and contained a provision authorizing the National Science Foundation (NSF) to provide grants for basic research in advanced information and communications technologies. Areas of research included affordable broadband access, including wireless technologies. P.L. 110-69 also directs NSF to develop a plan that describes the current status of broadband access for scientific research purposes.

Legislation in the 111th Congress

In the 111th Congress, legislation was introduced that sought to provide financial assistance for broadband deployment. Of particular note, provisions in the American Recovery and Reinvestment Act of 2009 (P.L. 111-5) provided grants and loans to support broadband access and adoption in unserved and underserved areas.

---

[50] The Joint Board recommended that the definition of those services that qualify for universal service support be expanded and that the nation's communications goals include the universal availability of: mobility services (i.e., wireless voice); broadband Internet services; and voice services at affordable and comparable rates for all rural and non-rural areas. For a copy of this recommendation see http://hraunfoss.fcc.gov/edocs_public/attachmatch/FCC-07J-4A1.pdf.

[51] For a detailed discussion of this Order and USF transition see CRS Report R42524, *Rural Broadband: The Roles of the Rural Utilities Service and the Universal Service Fund,* by Angele A. Gilroy and Lennard G. Kruger.

P.L. 111-5: The American Recovery and Reinvestment Act of 2009

On February 17, 2009, President Obama signed P.L. 111-5, the American Recovery and Reinvestment Act (ARRA). Broadband provisions of the ARRA provided a total of **$7.2 billion**, for broadband grants, loans, and loan/grant combinations. The total consisted of $4.7 billion to NTIA/DOC for a newly established Broadband Technology Opportunities Program (grants) and $2.5 billion to the RUS/USDA Broadband Initiatives Program (grants, loans, and grant/loan combinations).[52]

Regarding the $2.5 billion to RUS/USDA broadband programs, the ARRA specified that at least 75% of the area to be served by a project receiving funds shall be in a rural area without sufficient access to high speed broadband service to facilitate economic development, as determined by the Secretary of Agriculture. Priority was given to projects that provide service to the most rural residents that do not have access to broadband services. Priority was also given to borrowers and former borrowers of rural telephone loans.

Of the $4.7 billion appropriated to NTIA:

- $4.35 billion was directed to a competitive broadband grant program, of which not less than $200 million shall be available for competitive grants for expanding public computer center capacity (including at community colleges and public libraries); not less than $250 million to encourage sustainable adoption of broadband service; and $10 million transferred to the Department of Commerce Office of Inspector General for audits and oversight; and

- $350 million was directed for funding the Broadband Data Improvement Act (P.L. 110-385) and for the purpose of developing and maintaining a broadband inventory map, which shall be made accessible to the public no later than two years after enactment. Funds deemed necessary and appropriate by the Secretary of Commerce may be transferred to the FCC for the purposes of developing a national broadband plan, which shall be completed one year after enactment.

Final BTOP and BIP program awards were announced by September 30, 2010. For more information on implementation of the broadband provisions of the ARRA, see CRS Report R40436, *Broadband Infrastructure Programs in the American Recovery and Reinvestment Act*, by Lennard G. Kruger. For information on the distribution and oversight of ARRA broadband grants and loans, see CRS Report R41775, *Background and Issues for Congressional Oversight of ARRA Broadband Awards*, by Lennard G. Kruger.

Other Broadband Legislation in the 111th Congress

P.L. 111-8 (H.R. 1105). Omnibus Appropriations Act, 2009. Appropriates to RUS/USDA $15.619 million to support a loan level of $400.487 million for the Rural Broadband Access Loan and Loan Guarantee Program, and $13.406 million for the Community Connect Grant Program. To the FCC, designates not less than $3 million to establish and administer a State Broadband Data and Development matching grants program for state-level broadband demand aggregation

---

[52] For information on existing broadband programs at RUS, see CRS Report RL33816, *Broadband Loan and Grant Programs in the USDA's Rural Utilities Service*, by Lennard G. Kruger.

activities and creation of geographic inventory maps of broadband service to identify gaps in service and provide a baseline assessment of statewide broadband deployment. Passed House February 25, 2009. Passed Senate March 10, 2009. Signed by President, March 12, 2009.

P.L. 111-32 (H.R. 2346). Supplemental Appropriations Act, 2009. Provides not less than $3 million to the FCC to develop a national broadband plan pursuant to the American Recovery and Reinvestment Act of 2009. Introduced May 12, 2009; referred to Committee on Appropriations. Passed House May 14, 2009; passed Senate May 21, 2009. Signed by President, June 24, 2009.

P.L. 111-80 (H.R. 2997). Agriculture, Rural Development, Food and Drug Administration, and Related Agencies Appropriations Act, 2010. For Rural Utilities Service, U.S. Department of Agriculture, provides $28.96 million to support a loan level of $400 million for the broadband loan program, and $17.97 million for broadband community connect grants. Introduced June 23, 2009; referred to Committee on Appropriations. Reported by Committee on Appropriations June 23, 2009. Passed House July 9, 2009. Passed Senate August 4, 2009. Conference Report (H.Rept. 111-279) printed September 30, 2009. Signed by President October 21, 2009.

H.R. 691 (Meeks). Broadband Access Equality Act of 2009. Amends the Internal Revenue Code of 1986 to provide credit against income tax for businesses furnishing broadband services to underserved and rural areas. Introduced January 26, 2009; referred to Committee on Ways and Means.

H.R. 760 (Eshoo). Advanced Broadband Infrastructure Bond Initiative of 2009. Amends the Internal Revenue Code of 1986 to provide an income tax credit to holders of bonds financing new advanced broadband infrastructure. Introduced January 28, 2009; referred to Committee on Ways and Means and in addition to Committee on Energy and Commerce.

H.R. 2428 (Eshoo). Broadband Conduit Deployment Act of 2009. Directs the Secretary of Transportation to require that broadband conduit be installed as part of certain highway construction projects. Introduced May 14, 2009; referred to Committee on Transportation and Infrastructure.

H.R. 2521 (DeLauro). National Infrastructure Development Bank Act of 2009. Establishes a National Infrastructure Development Bank to finance infrastructure projects, including broadband and telecommunications projects. Introduced May 20, 2009; referred to Committee on Energy and Commerce and in addition to the Committees on Transportation and Infrastructure, and on Financial Services.

H.R. 3101 (Markey). Twenty-first Century Communications and Video Accessibility Act of 2009. Ensures that individuals with disabilities have access to emerging Internet Protocol-based communication and video program technologies in the 21st century. Introduced June 26, 2009; referred to Committee on Energy and Commerce.

H.R. 3413 (Capito). Rural Information Technology Investment Act. Authorizes the National Telecommunications and Information Administration of the Department of Commerce to make grants for the establishment of information technology centers in rural areas. Introduced July 30, 2009; referred to Committee on Energy and Commerce.

H.R. 3646 (Matsui). Broadband Affordability Act of 2009. Amends the Communications Act of 1934 to establish a Lifeline Assistance Program for universal broadband adoption. Introduced September 24, 2009; referred to Committee on Energy and Commerce.

H.R. 4545 (Murphy). Rural Broadband Initiative Act of 2010. Establishes an Office of Rural Broadband Initiatives in the Department of Agriculture which would administer the RUS broadband loan and grant programs, and would develop a comprehensive rural broadband strategy. Establishes a National Rural Broadband Innovation Fund, authorized at $20 million for each of fiscal years 2008 through 2012, that would fund experimental and pilot rural broadband projects. Introduced January 27, 2010; referred to Committee on Agriculture and in addition to the Committee on Energy and Commerce.

H.R. 4619 (Markey). E-Rate 2.0 Act of 2010. Amends the Communications Act of 1934 to create a pilot program to bridge the digital divide by providing vouchers for broadband service to eligible students, to increase access to advanced telecommunications and information services for community colleges and head start programs, and to establish a pilot program for discounted electronic books. Introduced February 9, 2010; referred to Committee on Energy and Commerce.

H.R. 5828 (Boucher). Universal Service Reform Act of 2010. Reforms the universal service provisions of the Communications Act of 1934 and other purposes. Introduced July 22, 2010; referred to the Committee on Energy and Commerce.

S. 1266 (Klobuchar). Broadband Conduit Deployment Act of 2009. Directs the Secretary of Transportation to require that broadband conduit be installed as part of certain highway construction projects. Introduced June 15, 2009; referred to Committee on Environment and Public Works.

S. 1447 (Hutchison). Connecting America Act of 2009. Provides broadband Internet investment tax credits and credits to holders of broadband bonds. Also establishes an Office of National Broadband Strategy in the National Telecommunications and Information Administration and provides broadband adoption incentives in telehealth and distance learning programs. Introduced July 14, 2009; referred to Committee on Finance.

S. 2879 (Rockefeller). Broadband Opportunity and Affordability Act. Directs the FCC to conduct a pilot program expanding the Lifeline Program to include broadband service. Also directs the FCC to prepare a report exploring whether the Link Up program should be expanded to include computer ownership in order to reduce the cost of initiating broadband service. Introduced December 11, 2009; referred to Committee on Commerce, Science, and Transportation.

S. 2880 (Gillibrand). Rural Broadband Initiative Act of 2009. Establishes an Office of Rural Broadband Initiatives in the Department of Agriculture which would administer the RUS broadband loan and grant programs, and would develop a comprehensive rural broadband strategy. Establishes a National Rural Broadband Innovation Fund, authorized at $20 million for each of fiscal years 2008 through 2012, that would fund experimental and pilot rural broadband projects. Introduced December 14, 2009; referred to Committee on Agriculture, Nutrition, and Forestry.

S. 3110 (Klobuchar). Broadband Service Consumer Protection Act. Seeks to improve consumer protection for purchasers of broadband services by requiring consistent use of broadband service terminology by providers, and requiring clear and conspicuous disclosure to consumers about the actual broadband speed that may reasonably be expected. Introduced March 15, 2010; referred to Committee on Commerce, Science, and Transportation.

S. 3506 (Landrieu). Small Business Broadband and Emerging Information Technology Enhancement Act of 2010. Seeks to improve certain programs of the Small Business Administration to better assist small business customers in accessing broadband technology. Introduced June 17, 2010; referred to Committee on Small Business and Entrepreneurship.

S. 3606 (Kohl). Agriculture, Rural Development, Food and Drug Administration, and Related Agencies Appropriations Act, 2011. For Rural Utilities Service, U.S. Department of Agriculture, provides $22.3 million to support a loan level of $400 million for the broadband loan program, and $17.97 million for broadband community connect grants. Introduced July 15, 2010; referred to Committee on Appropriations. Reported by Committee on Appropriations July 15, 2010 (S.Rept. 111-221), and placed on Senate Legislative Calendar.

S. 3636 (Mikulski). Commerce, Justice, Science, and Related Agencies Appropriations Act, 2011. For FY2011, provides $16 million to NTIA for the administration of BTOP grants and for the development and maintenance of the national broadband map. Introduced July 22, 2010. Reported (S.Rept. 111-229) by Committee on Appropriations July 22, 2010, and placed on Senate Legislative Calendar.

S. 3710 (Murray). Broadband Program Reauthorization Act of 2010. Extends authorization for broadband stimulus programs (BTOP and BIP) at $2 billion each for FY2011 and at such sums as may be necessary for each fiscal year thereafter. Introduced August 5, 2010; referred to Committee on Finance.

S. 3787 (Gillibrand). Upstate Works Act. Provides tax credits to expand broadband service in rural areas. Introduced September 15, 2010; referred to Committee on Finance.

S. 3967 (Landrieu). Small Business Investment and Innovation Act of 2010. Establishes a broadband and emerging information technology coordinator at the Small Business Administration. Introduced November 18, 2010; referred to Committee on Small Business and Entrepreneurship.

S. 3995 (Snowe). Federal Wi-Net Act. Directs the Administrator of the General Services Administration to install Wi-Fi hotspots and wireless neutral host systems in all federal buildings. Introduced December 1, 2010; referred to Committee on Environment and Public Works.

Legislation in the 112th Congress

The 112th Congress is likely to examine the efficacy of federal broadband assistance programs and how they may fit into the context of a national broadband policy. The following is a listing of broadband legislation directly related to the issue of federal assistance for broadband deployment in unserved areas.[53]

---

[53] For information on public safety wireless broadband legislation, see CRS Report R41842, *Funding Emergency Communications: Technology and Policy Considerations*, by Linda K. Moore.

P.L. 112-10 (H.R. 1473). Department of Defense and Full-Year Continuing Appropriations Act, 2011. Rescinds existing unobligated past-year funding for the Rural Broadband Access Loan and Loan Guarantee Program and the Community Connect Grants at the Rural Utilities Service. For FY2011, appropriates $22.3 million to the Rural Broadband Access Loan and Loan Guarantee Program for the cost of broadband loans, and $13.4 million to Community Connect Grants. Signed by President, April 15, 2011.

P.L. 112-55 (H.R. 2112). Consolidated and Further Continuing Appropriations Act, 2012. Provides FY2012 appropriations for Rural Utilities Service broadband loan program and broadband community connect grants: $6 million for the broadband loan program (subsidizing a loan level of $212 million) and $10.372 million for Community Connect grants. Introduced June 3, 2011; referred to Committee on Appropriations. Reported by Committee on Appropriations June 3, 2011 (H.Rept. 112-101). Passed House June 16, 2011. Reported by Senate Appropriations Committee September 7, 2011 (S.Rept. 112-73). Signed by President, November 18, 2011.

H.R. 1083 (Owens). Rural Broadband Initiative Act. Establishes an Office of Rural Broadband Initiatives in the Department of Agriculture which would administer the RUS broadband loan and grant programs, and would develop a comprehensive rural broadband strategy. Introduced March 15, 2011; referred to Committee on Agriculture and in addition to the Committee on Energy and Commerce.

H.R. 1343 (Bass). To return unused or reclaimed funds made available for broadband awards in the American Recovery and Reinvestment Act of 2009 to the Treasury of the United States. Introduced April 4, 2011; referred to Committee on Energy and Commerce and to Committee on Agriculture. Reported (amended) by the Committee on Energy and Commerce (H.Rept. 112-228) on September 29, 2011. Passed House October 5, 2011. Referred to Senate Committee on Commerce, Science and Transportation October 6, 2011.

H.R. 1695 (Eshoo). Broadband Conduit Deployment Act of 2011. Directs the Secretary of Transportation to require that broadband conduit be installed as part of certain highway construction projects. Introduced May 3, 2011; referred to Committee on Transportation and Infrastructure.

H.R. 2163 (Matsui). Broadband Affordability Act of 2011. Amends the Communications Act of 1934 to establish a Lifeline Assistance Program for universal broadband adoption. Introduced June 14, 2011; referred to Committee on Energy and Commerce.

H.R. 5973 (Kingston). Agriculture, Rural Development, Food and Drug Administration, and Related Agencies Appropriations Act, 2013. For Rural Utilities Service, U.S. Department of Agriculture, provides $2 million to support a loan level of $21 million for the broadband loan program, and $10 million for broadband community connect grants. Introduced June 20, 2012; referred to Committee on Appropriations. Reported by Committee on Appropriations June 20, 2012.

H.R. 6083 (Lucas). Federal Agriculture Reform and Risk Management Act of 2012. Reauthorizes rural broadband loan program at $25 million per year through FY2017. Introduced July 9, 2012; referred to Committee on Agriculture. Ordered to be reported by committee July 11, 2012.

S. 257 (Landrieu). Small Business Broadband and Emerging Information Technology Enhancement Act of 2011. Seeks to improve certain programs of the Small Business

Administration to better assist small business customers in accessing broadband technology. Introduced February 2, 2011; referred to Committee on Small Business and Entrepreneurship.

S. 1659 (Ayotte). To return unused or reclaimed funds made available for broadband awards in the American Recovery and Reinvestment Act of 2009 to the Treasury of the United States. Introduced October 5, 2011; referred to Committee on Commerce, Science and Transportation.

S. 1939 (Klobuchar). Broadband Conduit Deployment Act of 2011. Directs the Secretary of Transportation to require that broadband conduit be installed as part of certain highway construction projects. Introduced December 1, 2011; referred to Committee on Environment and Public Works.

S. 2375 (Kohl). Agriculture, Rural Development, Food and Drug Administration, and Related Agencies Appropriations Act, 2013. For Rural Utilities Service, U.S. Department of Agriculture, provides $6 million to support a loan level of $63 million for the broadband loan program, and $10 million for broadband community connect grants. Introduced April 26, 2012; referred to Committee on Appropriations. Reported by Committee on Appropriations April 26, 2012.

S. 3240 (Stabenow). Agriculture Reform, Food, and Jobs Act of 2012. Authorizes broadband loan and grant program at $50 million per year through FY2017. Introduced May 24, 2012; referred to Committee on Agriculture, Nutrition and Forestry. Reported to Senate May 24, 2012. Passed Senate (amended) June 21, 2012.

S. 3439 (Snowe). Federal Wi-Net Act. Directs the Administrator of General Services to install Wi-Fi hotspots and wireless neutral host systems in all federal buildings in order to improve in-building wireless communications coverage and commercial network capacity by offloading wireless traffic onto wireline broadband networks. Introduced July 25, 2012; referred to Committee on Environment and Public Works.

Concluding Observations

To the extent that the 112th Congress may consider various options for encouraging broadband deployment and adoption, a key issue is how to strike a balance between providing federal assistance for unserved and underserved areas where the private sector may not be providing acceptable levels of broadband service, while at the same time minimizing any deleterious effects that government intervention in the marketplace may have on competition and private sector investment.

In addition to loans, loan guarantees, and grants for broadband infrastructure deployment, a wide array of policy instruments are available to policymakers, including universal service reform, tax incentives to encourage private sector deployment, broadband bonds, demand-side incentives (such as assistance to low income families for purchasing computers), regulatory and deregulatory measures, and spectrum policy to spur roll-out of wireless broadband services. In assessing federal incentives for broadband deployment, the 112th Congress may consider the appropriate mix of broadband deployment incentives to create jobs in the short and long term, the extent to which incentives should target next-generation broadband technologies, the extent to which "underserved" areas with existing broadband providers should receive federal assistance, and whether broadband stimulus projects are being efficiently managed and how they may fit into the context of overall goals for a national broadband policy.

Table 4. Selected Federal Domestic Assistance Programs Related to Broadband and Telecommunications Development

Program	Agency	Description	Funding Amount (est. FY2012 unless otherwise noted)	Web Links
Broadband Infrastructure Deployment Programs				
Broadband Technology Opportunities Program (BTOP)	National Telecommunications and Information Administration, Dept. of Commerce	Provides competitive grants to public and private sector entities in order to provide broadband access in unserved and underserved areas; provide broadband support and services to strategic institutions; improve broadband access by public safety agencies; and stimulate broadband demand, economic growth, and job creation.	$4.35 billion (ARRA, P.L. 111-5) (2009)	http://www.ntia.doc.gov/broadbandgrants/
Broadband Initiatives Program (BIP)	Rural Utilities Service, U.S. Dept. of Agriculture	Provides competitive grants, loans, and loan/grant combinations to public and private sector entities in order to provide broadband access in unserved and underserved rural areas.	$2.5 billion for the cost of loans, grants, and loan/grant combinations (ARRA, P.L. 111-5) (2009)	http://www.rurdev.usda.gov/utp_bip.html
Rural Broadband Access Loan and Loan Guarantee Program	Rural Utilities Service, U.S. Dept. of Agriculture	Provides loan and loan guarantees for facilities and equipment providing broadband service in rural communities	$169 million for cost of money loans	http://www.rurdev.usda.gov/utp_farmbill.html
Community Connect Broadband Grants	Rural Utilities Service, U.S. Dept. of Agriculture	Provides grants to applicants proposing to provide broadband service on a "community-oriented connectivity" basis to rural communities of under 20,000 inhabitants.	$10 million	http://www.rurdev.usda.gov/utp_commconnect.html

Program	Agency	Description	Funding Amount (est. FY2012 unless otherwise noted)	Web Links
Telecommunications Infrastructure Deployment Programs				
Telecommunications Infrastructure Loan Program	Rural Utilities Service, U.S. Dept. of Agriculture	Provides long-term direct and guaranteed loans to qualified organizations for the purpose of financing the improvement, expansion, construction, acquisition, and operation of telephone lines, facilities, or systems to furnish and improve telecommunications service in rural areas. All facilities financed must be capable of supporting broadband services.	$145 million for hardship loans; $250 million for cost of money loans; and $295 million for FFB Treasury loans	http://www.rurdev.usda.gov/utp_infrastructure.html
Distance Learning and Telemedicine Loans and Grants	Rural Utilities Service, U.S. Dept. of Agriculture	Provides seed money to rural community facilities (e.g., schools, libraries, hospitals) for advanced telecommunications systems that can provide health care and educational benefits to rural areas	$21 million	http://www.rurdev.usda.gov/UTP_DLT.html
Universal Service High Cost Program	Federal Communications Commission	Provides funding to eligible telecommunications carriers to help pay for telecommunications services in high-cost, rural, and insular areas so that prices charged to customers are reasonably comparable across all regions of the nation.	$4.03 billion (2011)	http://www.usac.org/hc/
Universal Service Schools and Libraries Program (i.e., E-rate)	Federal Communications Commission	Provides discounts for affordable telecommunications and Internet access services to ensure that schools and libraries have access to affordable telecommunications and information services.	$2.23 billion (2011)	http://www.universalservice.org/sl/
Universal Service Rural Health Care Pilot Program	Federal Communications Commission	Provides funds to cover 85% of the cost of constructing statewide or regional broadband telehealth networks and of connecting those projects to dedicated nationwide broadband telehealth networks and the public Internet.	$81.5 million (2011)	http://www.usac.org/rhc-pilot-program/

Program	Agency	Description	Funding Amount (est. FY2012 unless otherwise noted)	Web Links
Appalachian Area Development Program	Appalachian Regional Commission	Project grants to support self-sustaining economic development in the region's most distressed counties and areas. Includes funds for a Telecommunications Initiative involving projects that enable communities to capitalize on broadband access.	$56 million	http://www.arc.gov/index.do?nodeId=21
States' Economic Development Assistance Program	Delta Regional Authority	Grants for self-sustaining economic development projects of eight states in Mississippi Delta region.	$11 million (2011)	http://grants.dra.gov/
Investments for Public Works and Economic Development Facilities	Economic Development Administration, Dept. of Commerce	Provides funding for construction of infrastructure in areas that are not attractive to private investment; most funding is for water and sewer infrastructure but some has been designated for telecommunications and broadband projects.	$112 million	http://www.eda.gov/
Library Services and Technology Act Grants to States	Institute of Museum and Library Services, National Foundation on the Arts and the Humanities	Provides funds for a wide range of library services including installation of fiber and wireless networks that provide access to library resources and services.	$156 million	http://www.imls.gov/programs/programs.shtm
Native American Library Services	Institute of Museum and Library Services, National Foundation on the Arts and the Humanities	Grants to support library services including electronically linking libraries to networks.	$4 million	http://www.imls.gov/applicants/grants/nativeAmerican.shtm

Program	Agency	Description	Funding Amount (est. FY2012 unless otherwise noted)	Web Links
Programs Related to Applications of Broadband or Telecommunications Technology				
Choice Neighborhood Implementation Grants	Office of the Assistant Secretary for Public and Indian Housing and Office of Multifamily Housing Programs, Dept. of Housing and Urban Development	Helps communities transform neighborhoods by revitalizing severely distressed public and/or assisted housing. Grantees may use funds to provide unit-based broadband Internet connectivity.	$110 million	http://www.hud.gov/cn/
Special Education— Technology and Media Services for Individuals with Disabilities	Office of Special Education and Rehabilitative Services, Dept. of Education	Supports development and application of technology and education media activities for disabled children and adults	$30 million	http://www.ed.gov/about/offices/list/osers/index.html?src=mr/
Telehealth Network Grants	Health Resources and Services Administration, Department of Health and Human Services	Grants to develop sustainable telehealth programs and networks in rural and frontier areas, and in medically unserved areas and populations.	$6 million	http://www.hrsa.gov/telehealth/
Telehealth Resource Center Grant Program	Health Resources and Services Administration, Department of Health and Human Services	Provides grants that support establishment and development of telehealth resource centers to assist health care providers in the development of telehealth services, including decisions regarding the purchase of advanced telecommunications services.	$4 million	http://www.hrsa.gov/telehealth/
Licensure Portability Grant Program	Health Resources and Services Administration, Department of Health and Human Services	Provides support for state professional licensing boards to develop and implement state policies that will reduce statutory and regulatory barriers to telemedicine.	$0.35 million	http://www.hrsa.gov/telehealth/
NLM Extramural Programs	National Library of Medicine, National Institutes of Health, Department of Health and Human Services	Provides funds to train professional personnel; strengthen library and information services; facilitate access to and delivery of health science information; plan and develop advanced information networks; support certain kinds of biomedical publications; and conduct research in medical informatics and related sciences.	$62 million	http://www.nlm.nih.gov/ep/extramural.html

Program	Agency	Description	Funding Amount (est. FY2012 unless otherwise noted)	Web Links
National Environmental Information Exchange Network Grant Program	Environmental Protection Agency	Provides funding to states, territories, and federally recognized Indian Tribes to support the development of an Environmental Information Exchange Network, including broadband infrastructure.	$10 million	http://epa.gov/exchangenetwork/grants/

Source: Compiled by CRS from FY2013 budget documents, the Catalog of Federal Domestic Assistance, and grants.gov.

Author Contact Information

Lennard G. Kruger
Specialist in Science and Technology Policy
lkruger@crs.loc.gov, 7-7070

Angele A. Gilroy
Specialist in Telecommunications Policy
agilroy@crs.loc.gov, 7-7778

www.ingramcontent.com/pod-product-compliance
Lightning Source LLC
Chambersburg PA
CBHW081245180526
45171CB00005B/549